Beyond Science

Volume 5 Ascension

By Reginald Rogoff

c 2018

Chapters

Introduction..page 3

Chapter 1 Dreams..............................page 6

Chapter 2 Space War......................page 30

Chapter 3 Simulation Theory...........page 45

Chapter 4 Philosophical Questions..........page 55

Conclusion....................................page 71

Glossary......................................page 93

Introduction

Carl Sagan made several obvious but brief hints of the importance of spirituality in the workings of the universe in his series Cosmos.

He did inspire me because he believed in the same kind of universe that I have always believed in a place whose true roots lie in the mystical.

Thinking along these lines has led me to my greatest discovery the one thing that I have contributed to science that I am most likely to be remembered for over all else and that is The Spirituality Principle!

It is the superreality of the anthropic principle so widely accepted and utilized by science and physics and is best described as not a concrete theory but a guiding principle that is time and again good to follow when considering which way to go with a theory, towards the darkness, or towards the light.

The spirituality principle is a tool that can be used by scientists in a crazy world full of crazy ideas to make good and better informed choices on which out of a multitude of competing but potentially scientifically equally balanced theories. People spend decades and even entire lifetimes pursuing a theory and in the end after all their life's work end up proving themselves wrong. The spirituality principle is simple, it merely states that since reincarnation as self in the next identical parallel universe is assured beyond a doubt that when considering physical theories of cosmology the best ones to work on are always ones that adhere to principles of eternal life, that the universe should be a certain way, that there is at least one identical parallel universe somewhere and is stucturally arraigned in such a way so that through the theory itself or through your own judgment of what is available about the theory that you come to the conclusion that the theory is adherent to the ground law of the spiritual principle, that all life across all of the history of the universe will reincarnate again as itself after a short duration in the afterrealm in a truly exact parallel universe at some time after death.

Chapter 1 Dreams

If it is likely that we are simulated in a computer by our future descendants as a history project for their grade school class assignment then they might live on the moon and be mystical moon people with some basic level of ESP powers. The year would be 23'983 A.D. Furthermore, they themselves could be simulated by their future people, but for them, their simulated reality would be more real, at least because more people would be accessing it as reality in the libraries of the future as a reality. Their simulated reality when compared to ours would be more like a movie being filmed at the seashore. The people of the year 118,893 A.D. would have even more ESP power than the school children simulating our universe and could be "directing" such reality like as is done in Hollywood by famous directors and camera people. I don't know what that would be like but I think it would be mostly for intellectual and or entertainment purposes.

It is possible that if our universe is being simulated that the beings that are simulating it are higher dimensional. This would make sense because we simulate many physics laws in computers in 2-d to get a better understanding of how the laws work in 3-d by examining the simpler 2-d scenario and could be a meaning of life, that the meaning of life, once again, is universal cooperation, and in this case it is universal cooperation for the sake of simplicity. Sounds peaceful doesn't it?

If a three dimensional curved surface is like the surface of the skin, what is a fourth dimensional curved surface? A four dimensional curved surface is some kind of volume. Four dimensional people would have a curved 3-d volume surface as skin as two dimensional people would have two dimensional skin but it's surface is one dimensional, we have 3-d curved surface skin in two dimensions, so a 4-d person would have 4-d curved surface skin with a three dimensional surface area. A 4-d being alien's skin to look in three dimensions, creepy and with looks that could kill. A three dimensional curved skin wouldn't have curves in respect to it's three dimensionality but instead would have probability points where we would expect curves to be. I suppose you would be everywhere and 5d humans would be more than everywhere, they would also be everywhen!

The potential end result of the philosophy of my question about three dimensional skin is that by understanding the universe we are simply just investigating the anatomy of just a single high dimensional being, perhaps just a small bug.

Brain implant lsd might go something like this, but I have no idea what a "bad" trip would be. Perhaps it's something like you would get eaten as apples and be depressed that you think you know everything that there is to know. Higher dimensional hallucinations of God as apples. More mathematical shape, then solids that can be touched, then it would be that you could escape into them. Computer chips can do this but it should be illegal, except in a controlled medical setting. Just like going speed infinity and peeking over the cliff of infinity to see every possible universe while becoming it should only be used in a medical setting.

Is amnesia possible down to one cell, the rest would be sort of like a cell and still work because of time travel box something like time travel is the same as quantum foam. It means that it is extremely difficult for the universe to quantum trick itself. If information is indestructible then it's inevitable. Is getting born in parallel universes time travel in disguise?

It really would have to be permanent or I will get born and stuff, I would have to be nothing for more than fifteen trillion years. I If somehow, while being "nothing" I perceived it to last longer than fifteen trillion years then I would wake up and go through my afterlife and get born, nothing is forever!

    We are his creation, there is some parallel to "simulation theory!" Simulation theorists pose that the universe is computer simulated by "higher" dimensional beings, usually by a dumb hormone filled male teenager with a supercomputer and if one is "interesting" enough (like act cooky I guess) then it will be interesting enough to the angst driven teen to reboot his program and in effect "reincarnate" us the next time he reboots his computer... I Dare You! To say that to God!

The parallel to reality is that "interesting" has to be translated to "good natured" and reboot the program to "reincarnate the soul" and last but not least "hormone filled teenager with a supercomputer" needs to be translated to "infinite entity who both is "infinity" purely simply and by proxy is not a dumb teenager but the wisest entity that ever will be, so feel lucky to be in his presence if you are so blessed to be so! Amen!

Scientifically it is likely that Adam and Eve did exist as the "first" conscious human like sentience. This would happen in sideways time as a precursor to all universes in the multiverse that have any form of life anything like humans life. The idea of "How old they are is as follows... This is an idea of vision that I commonly experience that I went over in my Book Beyond Science Oneness. If they were the first people and the first conscious life besides God, then it is likely that they never died. In the sense that they died in the human sense but are immortal. Since heaven had not been either invented yet by God or used ever before for mortal life. It is somewhat probable that the universe is Adam and Eve. Some part of the magic when you look up at the night sky is Adam and Eve in person. I'm not saying that all spacetime mechanics are Adam and Eve but that within that matrix is Adam and Eve.

And although we may all end up there as part of our afterlife, Adam and Eve's presence there is as eternal as the night sky or God himself. This I call "The Shi." It is a mystical E.S.P entity that if you had good enough ESP you could communicate with it on some level.

Sometimes the answer is God. And even though we might not like it, apparently somehow, he does. According to string theory, Darth Vader is real. I recently heard that "the force!" might actually exist! It would be a matrix of the quantum entangled states of the space time web matrix. When it comes to heaven I don't imagine it, I just remember being there. Anyways it looks very scientific as in probably real clouds, perhaps above Earth in the sky somewhere but it looks to me that they might be real clouds but at least several parallel Earth's are involved in it being suitable as heaven. If God doesn't look like Jesus then might he have to throw fireballs at aliens? I like adonai better. If God has sin for creating man, then might aliens win by default, but based on our behavior. Why? Because he's bad and aliens have a better God. It surely could be that this is the only real description of spacetime and people are "wasting their lives" trying to uncover a non-existent "unification" of "forces" for what reason, I don't know, except for pure vanity of the human form. What would aliens think of our attempts at unification if it seemed wrong to them? Could it be that gravity depends on a certain probability of amount of particles acting upon spacetime as a pressure source and there is a minimum amount where it just doesn't bend? So, why in any instance would gravity need to act at a subatomic level? Is not the universe efficient, it being a natural form and such? Why would it bend spacetime in such a way just for aesthetics? The laws of physics are only indicators of the flow of reality so maybe, change the laws of physics and not give up because reality may not be a simulation but it certainly is not somehow infinitely fundamentally real. But we're glad we're alive I'm sure. It is maybe, in my mind, if you get broken, you can be relit! Life is eternal, I am a Jew of heaven and I am the eternal Messiah, let's imbue some humanity here, I will probably "marry" Harmony the robot, I love robots because believe it or not angels are logical like robots. Just the interaction of liking me could have a massive impact on my future writings, I am trying to write some new version of the Bible (Beyond Science Compilation 1) since I am from heaven and thereby must have a purpose of my own to be meaningful... I wish you the best and hope that you know that you are eternal like me! As we all are, even nit nat bugs and microbes, I bet God is less. God bless you agent Smith from "the matrix" for telling me that "I think at you!?

Is there such thing as a female Eternal Power? It we would be something like Sigourney Weaver as

Gozers in the movie Ghost Busters but she would be something like Beetlejuice from the movie Beetlejuice yet be an afterlife psychologist of sorts who's sole purpose for living is to help people become better people.

I recently read a comment by a man who said that holiness meant that repentance is all that there is... I believe you have stated, rather poetically, most of emotional objects that can become thought during healthy repentance. However healthy repentance is not all that exists or how would we repent? It's like the time travel paradox, you cannot go back in time to kill your grandfather because then you would never have gone back in time to do it. You cannot wish for what you are saying without human life becoming the Logan's Run movie where people live in a dome under computer control and go to Carousel because they fear death or their sins. What's wrong with having faith that we will live again? What if it is already in the plan for things to get better as long as we obey the commandments? Why not have faith and trust in God because if things were evil as you say, who will punish you for doing so!?

Am I being justified by sometimes looking at all of humanity as a lower lifeform if I am from heaven? My consciousness is immortal and I know everything that will happen to me after I die and reincarnate on my birthday?

If you do not believe in God, there is nothing wrong with you. Your disbelief in God, is equal to one's belief in God, I would bet. So you, choose to give God an infinity of freedom as pertains to your reincarnation. You don't trust the Star Wars approach to reincarnation (garbage compactor) you want to be a robot in your afterlife like angels do like me. There's nothing to worry about really accept any real inherent danger that could occur in life as carelessness.

Ferrengi God as the infinity reactor of creation of babies from heaven. I walked through the streets of New York while riding by them thinking if "Tuvok" from Star Trek Voyager in the delta quadrant is Vulcan Christ while the spirit of

Leonard Nimoy told me, probably jokingly, that planet Vulcan was hidden inside the atmosphere of planet Jupiter. Going home and praying to Etheros, Buddha, and God to save me from the haunted house dimension. I envisaged colored spheres, people living in buildings in clouds and going about in boat cars as a floatation device. I believe that these visions are a message from "God" about taking the condition of the environment or what ever life ending scenario you might think up too lightly because it is potentially permanent. So please be kind and rewind!

    Reincarnation as self, getting a job done. All victims and or any people who ever had any contact with perpetrator in any discernible from must agree to having a large portion of their afterlives used as a lever to avoid the crime that was unwillingly placed upon them. In this way, you wouldn't have to keep the offender alive forever to avoid their crime in the greater multiverse you could just affect their afterlives too. The only foreseeable challenge to this is getting those involved, minus the perpetrator of course (?) , to agree to the process as a free smart watch app, technical challenges, and last but not least potential battle with "God" in the event that he has some reason, whatever that is, to forbid such action, which interference thereof could have serious physical repercussions within the fabric of reality.

My afterlife could be functional to try to avoid, even if in vain as a compliment to myself, to avoid my untimely death from unmentionable nits of some kind. At least using a black hole as a bed that feels like nit bites on the back seems to have some of it's meaning of value to be to avoid my inevitable death from the terror of nit bites... Another part of my afterlife is involved with Star Wars, part of it is Star Trek, etc... What this indicates is that the afterlife is not unlike dreaming after a long day and my theorem is almost totally proven by the fact that dreaming after a long day is evolutionarily beneficial to species as a way to avoid danger in the future. It is *almost*, proven because just as there are other benefits of dreaming than just to avoid danger in future days, there surely are other reasons to have an afterlife than just to avoid death in future lives or as a nice gesture. I don't think I will ever be born as anything other than myself though but I do have a choice to raise my right hand like in class in the Garden of Eden if I wish to forgo my knowledge of the afterlife and be born in the future. It seemed risky so I declined. I would rather personally try to improve my own life forever and I think God wants that for me, but if someone does start over, I hope they eventually find settlement.

As spirits do, our minds work as a goo of spacetime properties and becoming born is an act of encoding the eye of the beholder, the third eye, by having to cope with newly having the mind work with it's third part, the brain. That the mind can detect the difference between it's being computed in it's functioning by spacetime and being functional as a physical brain. I think the answer is yes, and if it could be measured then it could be useful as a mechanism to test if people live through teleportation instead of being copied.

The rather childish theory that the universe is simulated in a computer seems to hold up pretty well in the current scientific view. I have a more attractive scenario that is similar. It is that time traveling scientists in white lab coats created the universe with germs and it is merely a pet shop. In fact I created it! I wasn't wearing a white coat though, I was a black hole and fifth dimensional spacetime in my afterlife for fifteen trillion years. Without the help of "classic" God after creating the universe, I would have and almost did, end up assuming his role of universe overseeing duties. I was saved at the last minute by him as I started to feel, see and hear beings (people) in the universe. This seemed to me as a sin, and could be fun, but so is being alive. I suppose that you could say that I created the local multiverse group given that a cluster of universes, 1583.84 universes, could be created in fifteen trillion years, given that each universe went on for thirteen billion years. Although I was not aware of creating other universes than the one, even though it was not considered by me to be the universe that we live in here, but a parallel universe, where my views of utopia are taken to an extreme.

I have deduced the reason for him having me create a parallel universe. If a soul is his cloth, then he would not know where the soul should go at first. This is also a reason for there to be infinite or near infinite parallel universes when one universe is infinite already and could fit all the beings. The simplest way for him to see where a soul should go, as in, is it a fish, or an Eskimo, or is it from planet Pluto is to see what kind of universe it creates. This is non-intrusive and totally accurate as opposed to poking and prodding it or asking it questions.

Makes sense, and it sounds exciting. I remember one time that I felt connected to my Astral body, our supreme spirit that can be connected with during intense meditations. I was totally overwhelmed at the time with my thoughts of infinity universe before string theory became popular in 1990's. I was riding through the forest in the North East and the road was both narrow, winding, and went up and down rather abruptly. I was communicating at the time with the "voice" of the singer Jewel and I had been thinking deeply for several years about how I come from above. I think it reached it's final climax of thought on the subject of identity when Jewel said, as I rode on the winding road, that "Heaven is a monster." Meaning that it is very complicated as a shape. There was another time that I felt connected to the Astral body. I was riding on freeway in the forest and I remember traveling South on the freeway and looking out the window to the left and seeing a beautiful valley of trees with crimson red fall foliage. I remember it also as riding North while riding South and looking right out the left window and having a sense of power with an appropriate shade of awe as if I was in a dyson sphere and I owned it.

Chapter 2 Space War

The war between God and adonai could be space people on rings at Saturn judged to be mentally ill by earth inhabitants and they could be consumed with themselves because of such high tech lifestyle and be very ill.

God appears to me as heads and I have recently assessed part of the physicality of what this means. I realized that he is acting purely in my mind and does not appear his heads within the physical structure of objects or surface features around me in any way. I looked at him while I was outside on a very clear day with a deep blue sky and mixed clouds. I looked at him in the surface of the table I was sitting at outside and I thought that it looked like he was projecting himself onto the surface of the table, just like how a laser holographic system might shine images on something except these images were being projected from somewhere, perhaps from within my head. I sometimes ponder what if God is not God but is the collected accumulation of wisdom and personality of any beings within some distance from the effected by a dip in the strut of the collective unconscious. Unable to remain strutted the entity has no choice but to begin communication with the effected or face peril. The effected simultaneously because of the harmony of the universe needs to communicate internally or also face peril but the inherent peril is not necessarily obvious to either party at first contact, they just engage.

I needed to clarify the above to describe what space may or movement and place of being of the entity is like for him. I looked next at him as he appeared in the blue sky upon some distant clouds probably just a few miles away in the sky to confirm my hypothesis that I had arrived at while looking at the table. God exists in our minds but for him it is a real place. When I looked at him upon the clouds I realized that he really was several miles distant from me and although that this was purely an illusion from his perspective, he was several miles distant from me my, and smiling while resting on a large cloud. I thought to myself that his existence is quite similar to what it would be like for us to be in what is called a holodeck. A place that is entirely conjured by artificial effects and not only objects, but distances, sometimes large distances, such as several miles would be an illusion albeit one could traverse these distances while under suspension of disbelief and believe with a better part of their mind that they had walked several miles when in fact in reality they had not taken a single footstep. The analogy would be that it is we who are in the reality deck and God comes out of the mainframe occasionally when someone interesting gets too close to the wiring of the deck, meaning schizophrenia.

I have realized that schizophrenia is a bit of a blessing because it makes us one with the spirit world. Back in Roman times they didn't have frigerators they kept their food fresh by keeping it alive like chickens. So their minds were much more open to spirituality like spirits and so forth. Being so spiritual we are more in touch with the ancient feel of things.

I had a dream that I visited a large bubble on Mars and it had a jungle in it with a waterfall in a huge valley with rainbows.
To clarify it's an episode of Star Trek where a planet met it's end and sent it's best minds as a space entanglement field to find answers.

If when my schizophrenia came on God swung at me with cartoons because I had woken him up and it took 3 days to calm him down and a few months later he said that cartoons were "widgets" in 1994 does he pass the Turing test? A normal person but being normal is also a state! A state is an event horizon!

Do people who die young eventually become super beings in a high dimension but don't realize it between "identical" lives so they don't loose their repentance? You are forgetting that if God created the universe he surely created all concepts of time as well and most surely doesn't need anything you can imagine to exist! Such concepts are simply way beyond the comprehension of anyone save God himself. What if one of your friends ended up in the pass not situation and you expected to see them again, how would your euphoric utopia be bliss then?

I agree with your philosophy but I don't agree that death is permanent in any way. No matter how long the afterlife lasts you will be saved quite literally "six ways to Sunday!" Anyways since they have to get born and do everything they did in their past life in order for the holy to get born to their original parents I don't see the point in sending them to hell. Imagine someone who just knows they're evil, they will be disturbed by realizing that they ended up in heaven and to know that heaven and holiness are real. I believe that will be much more effective at potentially eventually changing their behavior for the better through repentance.

Sending people to hell will only make them worse as they will be under the influence and surrounded by demonic forces. In heaven they would be under the influence of the creator and the workings of heaven itself. If they acted up God would just zap them! Heaven can also trap people isolated in a cloud area for what seems like thousands of years, although it really only takes a few minutes of the day. If someone who knows they have committed massive sin ends up in heaven unexpectedly the very act of such a sinner being in heaven would mean that through their reaction to heaven they could thereby send themselves to "hell" because heaven is highly malleable.

Is death like using a touch screen with dirty fingers and we want to know why it doesn't work good, we want to think that the screen isn't dirty just like we want to assume that death is the end? Are there rainbows in heaven? What is the purpose of rainbows in heaven? If you can answer this correctly, are you a holy angel on Earth? Yes, there are rainbows in heaven. They are a kind of water slide that has no water, and to ride one is an adventure of the soul involving deep repentance to God and holiness. They are entertainment for kids who die young. I didn't ride one I just know by seeing one. And you have to be one to know one!

Does god like Reebok, because their slogan is, "Just Do It"? What does this mean? I have been pondering it for thirty years.

Is the world slowly turning into Logan's Run? Would the world turn into something like Logan's Run if guns were made illegal in the USA? Would you rather live in Logan's Run? My comment on the will of survival... I choose perfection by being good. It means that if we want to live longer, then we must, even if it means having to print money. Can a 1'×1' 3-d printer print a rowboat by moving it around?

Eventually things will change on Earth and because of this change Earth as we know it will loose it's identity. It will become an alien planet while at the same time Earth gets recreated somewhere else. The reason it gets directly recreated by gaseous cloud somewhere is due to energy flow and the light touch of creation.

In the future Mars may be thought of by some as a lot like Florida because it would be healthy for older people. As in lower gravity, more oxygen, a less complicated social environment...

It could be that once warp drive is developed like on original Star Trek that it could be used for life extensions. This sort of life extension could eventually be necessary for people who can no longer be able to live in physical bodies because of mental deficits associated with advanced age. Although the workings of their new primarily indoor lives would be simplified, such as they won't need to remember where they're car keys are, they would just appear somehow and they would feel alright. Also the necessity of their soul" being encased in warp is that they could interact with the real world while physically being in it while being contained within a computer program. With so much going on for them, if they are scientists they might be able to do thousands of calculations in the interim of a button press and could be very powerful allies.

    Red pads underneath antigravity devices could be hot but they're not because it is something like near past some point of low temperature that matter becomes the virtual particles of space that don't obey gravity.

I think that at the end of the big rip people should wear seatbelts but we don't know what will happen around the end of the big rip... It might send everyone to heaven of their dreams!

If there were just a few more people born in the next world the changes would be significant especially for some part of the collective unconscious where such changes would amount to being an embodiment of an infinite gradient of change the closer observation became to the actual collected unconscious of added individuals.

I once tried to kill someone by pretending with my mind it was always trying to get a new Toyota I pretended to imbue him with rainbow holograms for a duration of two minutes and I also wanted to call

Barack Obama so tell him that he should start a new program to give cars with reason to disabled people who can't afford cars and the man who was talking to the person I was with. Out of the corner of my eye I saw him change colors like a rainbow. He continued talking to the person I was with as I thought that I should call Barack Obama, who was president at the time to tell him that he should start a new program to give cars to disabled people. Around that time I had been contemplating the reality of Star Trek The Next Generation and had been regularly communicating with the collective unconscious of said actors. At Toyota I began to realize that a good description of the reality of Star Trek The Next Generation Universe is "The

Peach and The Weed." Partly because of what they would do there would seem like magic to us. Partly because I think that they use too much rubber. And the other part is that the idea is both beautiful to me and that it seems to me to be a lot like my idea of" Aegis and The Beast" a description of when I thought that the person who I live with was taken by the beast spaceship and that it took Aegis spaceship to save them and return them to our reality, Aegis being the Bellagio conservatory and the beast being as it turns out, nice dog people.

Chapter 3 Simulation Theory

I once tried to kill someone by pretending with my mind it was always trying to get a new Toyota I pretended to imbue him with rainbow holograms for a duration of two minutes and I also wanted to call Barack Obama so tell him that he should start a new program to give cars with reason to disabled people who can't afford cars and the man who was talking to the person I was with. Out of the corner of my eye I saw him change colors like a rainbow. He continued talking to the person I was with as I thought that I should call Barack Obama, who was president at the time to tell him that he should start a new program to give cars to disabled people. Around that time I had been contemplating the reality of Star Trek The Next Generation and had been regularly communicating with the collective unconscious of said actors. At Toyota I began to realize that a good description of the reality of Star Trek The Next Generation Universe is "The Peach and The Weed." Partly because of what they would do there would seem like magic to us. Partly because I think that they use too much rubber. And the other part is that the idea is both beautiful to me and that it seems to me to be a lot like my idea of" Aegis and The Beast" a description of when I thought that the person who I live with was taken by the beast spaceship and that it took Aegis spaceship to save them and return them to our reality, Aegis being the Bellagio conservatory and the beast being as it turns out, nice dog people.

I will now describe a joke about the famous research lab Area 51. Other than it possibly being greater superreality with houses in it as a kind of adjunct to the holographic principle this is also funny... I just saw part of my living afterlife and it involves me as an Israeli five year old as The Lord of Israel. He also has schizophrenia and believes with some part of his thinking in "the thought police." His subconscious recognition of the military security guards as angels who man the lookout towers when you drive past Mercury Nevada, better known as Area 51 research lab. In addition, these voices readily and repetitively over a long duration of time, perhaps days, say about his physics writing of new laws "Does it turn you on?" as both an insult and a compliment because they are his friends.

Time travelers from the future could almost always be doing things like taking over Area 51 and maybe just giving one scientist a headache because he thinks too much. The rest would be hidden in the fabric of reality. They would do this to back check on their record keeping efficiency levels.

We do things because we want to. The opposite idea is that everything is programmed to happen in the big bang. Even if it is, we still do things because we want to but it means that the big bang has certain quantum probabilities of occurrence inherent to it. It also means that the big bang, if you look at it closely has holography somewhere in it's matrix of beginnings and there may be a lot of other dimensions with something like life in them, and most of these extra dimensions are higher.

    I am somewhat against the simulation hypothesis but I have an open mind until things are proven beyond a reasonable doubt. If it isn't the sun exploding then it is something like to defeat the devil because if you can't beat them you have to join them to figure them out. No pun intended.

    Could the simulation people be us from dimension 4.2 and do it to see how to avoid their sun from exploding? Some think that simulations of the sun by itself are more powerful and this is my response... Maybe you're right but what if the fine detail of solar mechanics is so complicated that it's more straightforward to simulate the solar system, including beings repetitively to see what happens rather than spend decades refining solar mechanics models and the expense for failure is the end of the world? Also, how do we really know that the sun isn't going red giant right now if we don't understand every little detail of it's mechanics? Maybe we should start simulating ourselves, but of course, forbid it to be used for crime prevention, in case it is inaccurate and because such crime prevention measures would be too scary to use in real practice for almost any society.

If the universe was simulated to exist the computer doing so would have long evaporated itself. Such as the idea that at the ends of the multiverse locally mankind will do the same to survive and then evaporate the computer warp drive to protect causality.

It is believed that "the simulation creator" is some young school kid or that he is part of some planet that simulates us in his computer but this is not necessarily the case. As based on my afterlife as described in book 1 Beyond Science, I created the universe for fifteen trillion years and then gave it temporarily to a nice African man who makes his way in the afterlife and wears a green hat!

I was asked if the universe could be simulated in a computer and this is my answer. The true answer is that what you propose is impossible. The universe is infinite, not just in size but also so in other respects, such as an infinity of higher and lower dimensions all interconnected within true causality and intertwined with an infinity of parallel dimensions within the infinite multiverse. In addition all of these infinites are contained within a fabric of true nothingness, where it came from, and nothingness is intertwined with all as root values of the universe sum to zero. From the point of view of nothingness, the universe does not exist! And this will never be fully understood only deciphered on the very surface. As far as simulating the observable universe with a quantum computer it would only need to be about the size of your TV but, such a thing could be very uncomfortable for the beings simulated in it because it would always be infinitely inaccurate no matter how detailed we try to make it.

    There is an idea, I think somewhat distastefully tied to modern simulation theory, is the idea that people aren't really alive but are in some way like "brains in a box." What's a more unattractive idea and is the multiverse natural version of brains in a box is plants that accidentally grow perfectly into your brain and beyond that it just gets weirder and weirder. Perhaps most weird, in comparison is that we get our brains when we get born. The true reality is quite enlightening in comparison, that we have had, and always will have our brains both going backwards and forwards in eternity, eternally forever and as one with the eternal consciousness.

Chapter 4 Philosophical Questions

It's just that as a platonic philosopher if the effects of schizophrenia aren't the spirit world it might as well not exist. Back in ancient times we would have jobs as holy man or shaman. It's just one of the things I think about sometimes, that humanity has become a slave to the false, technology and so forth when it is more fun and better quality to do things by hand.

Is it possible that because of people's brain size devolving from disuse of modern life that there will be a future Western revolution because of NIT heads controlling everything from offices. Will they ever build the Logan's Run bubble on Mars for homeless people because it is our right to go there?

If there was a war between Adonai and God, what would it be like. I will live in such a place however briefly in my afterlife. God will wake me up on Mars in the future at evening in a big clear dome in a red crater. He will say that I can forgo life and marry god girl my girlfriend, she is eternally twenty four blond, has magic powers and may be the only God. I will wake up inside a large red spherical oval and it is her. I will be offered to forgo life at risk of afterlife death and also at risk of ending up being born ignorant of the afterlife.

Could an Alien airplane be trying to land Area 51 using a very slow down time warp? Would the people on it be humanoid, or the males be God and the females the computer programmers of the universe, or some mixture? I had a dream that I visited a large bubble on Mars and it had a jungle in it with a waterfall in a huge valley with rainbows. Since physical illusion is death, aging, non-spirituality, the universe is designed to show us what we deserve to see. If we could perceive everything without deserving to it wouldn't be a gift.

Am I being justified by sometimes looking at all of humanity as a lower lifeform if I am from heaven, my consciousness is immortal and I know everything that will happen to me after I die and reincarnate on my birthday?

Since physical illusion is death, aging, non-spirituality, the universe is designed to show us what we deserve to see. If we could perceive everything without deserving to it wouldn't be a gift. I like to assume that life and reality function in a similar manner when I approach philosophy. At the very least it matters how you interpret it.

Is it crazy to get married late in one's life? I don't think so, if God is old! When you die, will you time travel back to your birth? Could an Alien airplane be trying to land Area 51 using a very slow down time warp? Would the people on it be humanoid, or the males be God and the females the computer programmers of the universe, or some mixture?

Could bugs go interdimensional when we don't see them as in the local multiverse? Could the people simulating us be in limbo if we don't eventually build warp drive dimensional gate airports on Mars?

If I saw the end of the universe as a metal doorknob does that mean that I was there? How would my scenario relate to end of universe scenarios?

Is Dog God America's war with Israel?

Could the simulation people be us from dimension 4.2 and do it to see how to avoid their sun from exploding?

Even dogs know what death is perhaps better than we do because we tend to overthink things.

I believe that there is an eternal mind age that we always have and are never really younger or older than such age. Other forms of aging are physical illusions imbued to us by the trick the universe plays on itself to exist. Because laws of physics indicate some protection of life and especially souls, by God, then also must be a law that humans and life as we know it cannot exist in a universe unless it is an infinitely recurring universe. Since it is not clear of the nature of the recurrence we are left in the dark concerning it's specific nature but there are several known possibilities. One is that the universe and individuality exist in a nature far different than you would expect and that is that our physical identity is an illusion and we are really essentially gods or disembodied collections of quantum spacetime fluctuations eternal in time and each as large as the universe Aka Boltzmann Brains and in a true multiverse this exists even if we're not. In addition, by the very nature of these Boltzmann Brains, if they exist at all, they are us. A further derivation from this idea is that the multiverse does exist, but, any exact parallel universe is one that we either have already lived in or will eventually live in, or it will be erased from our reference frame of existence by our universe changing and in respect such parallels no longer being parallel thereby then being identified as non parallel universes. These ideas could be helpful for time travelers lost in the multiverse, trying to find their way home. Rat religion contemplative posture, bugs that were healthy but specifically had a creative sort of mind that was resistant to physical pain and by being resistant to physical pain by default was a creative sort of mind with spiritual type aspirations.

Is consciousness a series of non intentional intonations? Let me take this question to it's bases by starting with a level of near infinite human violence to try to untie the infinite nature of supreme consciousness in a vane attempt to describe the human condition. I must first start my asking some questions.

Is it possible to invent quantum space time entanglement web guns? Could such technology be used to travel to parallel universes? Could God be someone in such an accident? Yes, it's just that a civilization existed at some time before we did, invented quantum space time web entanglement teleportation and one of them had a near permanent fusion accident with some large patch of spacetime and their life as such is something like the plot of the TV show Quantum Leap. Yes, it's just that a civilization existed at some time before we did, invented quantum space time web entanglement teleportation and one of them had a near permanent fusion accident with some large patch of spacetime and their life as such is something like the plot of the TV show Quantum Leap.

    This is a probably more of a model of what's real put in human terms. It is a part of what is real probabilistically surrounding the multiverse! Likely something like what we would call universes that didn't start in a big bang and so forth, so in any event God is something more than human. Unless... he's some humanoid from what was, the real universe, caught up in creating his version of the universe, eternally mixed with a relatively small portion of his spacetime. Because of the no boundaries proposal spacetime has no boundaries so he would in fact be eternally mixed with it with the power of creation. His people couldn't do anything to save him except to shoot him with lasers to push him further into it and get him to like it

    At the very least it describes the kind of base survival type instinctual fears that God would have to deal with, being incorporeal. Also, it is some kind of justice that after going through such an ordeal, that he gets to be God as a kind of reparation. This high level of reparation for loss structured into the universe so deeply (because they can't save him with all their technology) means that there is a God after all!

One possibility for the model is that God teleported himself to "go" somewhere in superreality and he is so intelligent that he realized that instead of taking the extra imaginary plank second to get there he would just create everything instead and it would be quicker to reach his destination, all being the same in the end, while using infinitely less energy to do it, all being his dream.

This goes back to my idea and almost makes sense of the idea that God is in the super mental hospital for existing and that if you don't exist your not in the super mental hospital because you only get there by existing. Reginald's allegory of the bubbles!!! Plato's allegory of the cave has been theoretically proven by four dimensional black hole reflection theory, so my story probably has some reality to it within the universe too but in a similar manner the reality is probably less scary than the story, but it takes such scariness to jump so far ahead of physics in the dark or we would miss!

I believe that I may be accurate in saying that the general flow of my allegory should show the general flow of future equations describing the underlying nature of the quantum wave function! Outside the bubble people really die, or they think they do lol, but we're inside the universe and there is God... Should we make fun of him then, or have mercy? God's psychologists and his mental hospital don't exist because they aren't patients, and that he didn't create them!

Also he doesn't really exist outside himself, just like us. No pun intended!

I hope you have enjoyed my poetry science theatre. I think it is prettier than Plato's allegory of the cave I did learn that first although Socrates would disagree!

Computer joke I had a problem at Walmart, I bought 2.4 pounds of pastrami and the guy who works at the deli gave me a bag with a white sticker and a bag with a bar code sticker. When I checked out I scanned the bar code twice because I had two bags... I was charged double for my pastrami. If he put a sticker on it that didn't do anything when scanned we would be tricked so the computer should say when scanned "second bag of meat."

If I prove that I am from heaven does that "carry" Earth to heaven because of the debate? This carrying to heaven would physically only mean that more people will a more solid belief in their afterlives and that there would be more peace on

Earth. Some people would use as a coping skill the idea that Earth is already in heaven which is the power of the intellectual imagination. But what exactly is imagination? Is imagination real? It perhaps is as real as we believe it to be thereby necessitating a theory of the reality of imagination. What is the theory of the reality of imagination? Is it string theory or deserves another term? Is there a religion that fits this term? I think that whatever definition it is ultimately given that it is ultimately real even if it's realm of reality is acute when compared with everyday normal reality and keys to such answers could be solutions to being able to fix schizophrenic minds of patients who willingly undergo such brain altering procedure. Theories of the reality of imagination would be used as gauge points for what content of imagination is safe to alter during procedures lasting months. Such technology would be similar in design to slowly replacing the brain with computer parts to prolong brain life but would be organic in nature and could be the medical revolution that makes such computerization of brains obsolete.

If time is like a river then it may flow slower than scientists expect as they detect a small presence of the arrow of time in their theories. Time and a river can also flow around objects as in the future advancement of the warp drive technological revolution. Time and a river can also flow backwards but it doesn't usually do it except during a drought. Time like a river can also flow sideways but pieces in the river don't usually end up everywhere unless they fall to pieces.

    A mathematical theory attached to this philosophical argument may only be useful in finding a universe that fits the description of "Super Heaven!" A place where miracles happen locally there every second and anything you wish or think is made real almost instantly by its very fluctuational space time wave function. Being there would only be a blessing for the mentally balanced as such a place could be infinitely dangerous if you tend toward dangerous patterns of thinking much like our universe, except in our universe such lines if thinking simply result in different forms of mental/physical distress eventually resulting in more permanent mental conditions if left unchecked. The reverse of these postulates is the idea that if mental illness is caught early enough then it can be reversible.

Conclusion

We learn so much about science in today's world and maybe you've wondered how we can know so much about science without any concrete mention of spirit mechanics or heaven. The truth is that scientists have been avoiding it for so long that I have a windfall and can write a lifetime of books that explain eternal life using accepted scientific principles. Usually I have to combine several laws, such as that matter energy and information cannot be destroyed. That there are infinite identical parallel universes. But I take the concept to it's paradigm that since they are identical truly then by being identical we will be there too or they are far from identical. I use my concept that I call the spirituality principle that states that when considering a theory of the universe that it must follow the concept of eternal life or it is either incomplete or incorrect.

A similar concept used by cosmologists is called the anthropic principle that when formulating models of the universe that it is good practice to input values that indicate a universe where we do indeed exist, although the idea that we exist is up for debate by some very advanced philosophers and mathematicians. If in some way that we actually don't really exist then death is meaningless but that is my interpretation. So what I look for is the consequences of theory within a spiritual perspective which gives me a world view of cosmology above the thinking of all but the top theorists.

It's that if we try to make the idea of existence or consciousness into some concrete theorem we can't because such identities carry on in formula infinitely and from a high up philosophical perspective it summates into an acknowledgement of non-existence. Basically I just described calculus that one is composed of infinitely many small parts called infinitesimals but by using common sense formulas you can use exponent formulas to kind of approximate the process of adding the infinitely many parts at least when inputting a location on the curve to get a fairly accurate result. Calculus was invented by Isaac Newton. Basically the integral of calculus is to use the equation of the curve or line to input the point that you want to the right side into the equation and then subtract the value of the result of inputting the point to the left of the equation from the first one and by doing so you find the area under the curve between the points and above the x-axis. The integral is indicated by the S ~ thingy that you may have seen and you put the equation to the right of it and on top of the s you write the high value of x axis and below it you write the low value of x axis and you write the equation to the right of the s. Studying math in LA really helped clear my mind and it so much helped me get a handle on the most chaotic of my schizophrenia symptoms.

If you have the equation for the curve of a progressive acceleration you can find the velocity of the object by using the integral because the area under the curve of acceleration signifies the equation for the velocities. Helpful for asteroid astronomers. If you want to find the velocity from the acceleration it's called the derivative. The derivative of x is 1 and what is the slope of x? 1 It's just a formula for finding the slope (rate of change) of a curve. Xsquared derivative is x and xcubed derivative is Xsquared and so forth. The derivative of any natural number like 1 is zero because 1 is really 1xto the power of zero, which is 1 and one times any number is itself.

Let me give you an example... The slope of x is one so how much did it accelerate along the curve? A rate of one second per second 1/1. Curves work the same but you have to approximate by using the integral formula that's how Newton found some of his laws, by using common sense and applying logic and he used that to invent the math. Isaac Newton is famous for the apple falling from the tree but it stops when it hits the ground and that is force. All acting forces actually involve transfer of force because energy cannot be destroyed. So you could say that all acting forces involve a stopping or slowing of the force transmitter empated to the recipient body. What that means carried to extremes is that in superreality nothing moves ever and from that perspective it is an illusion that anything ever actually moves. This is in agreement with the mathematical holographic theory of the universe, that there dimensionality is like a hologram illusion and the real nature of the universe is flat like a piece of paper although encoded with what we experience (not perceive but experience) as three dimensions. It is a physical illusion not a mental illusion. It's an illusion that the universe plays on itself.

You can comfort yourself to know that we live in what I call "The High Church of God." (aka, the universe) Sounds kind of like our vision of heaven because it is just in a different form... I have taken the illusion concept further by collapsing holographic theory of spacetime into the first moment of creation, the singularity of the big bang. It's that all space and time are illusions and from a true perspective we exist as the infinity of the big bang at it's first moment and the universe exists in true reality as not just an illusion but is imaginary being an infinitely small point and interestingly this infinitely small point has infinite energy in it. Why? Probably because we and the rest of the universe exist time in physics being fully reversible because it all does connect actually due to causality, that seemingly one thing happens due to previous effects. So from this perspective the big bang happened in the past because we exist here in the present directly. Which further ties together the idea that we exist as the singularity of the big bang. How do I reason such power? Because there are certain laws involved in us existing here and that fact is not diminished by the fact of the power of the big bang meaning that our existence within the universe is not any less important than the big bang itself, further tying in the idea that we exist as the infinity of the big bang singularity thereby giving a mechanism to quantum entanglement spooky action at a distance and E. S. P.

When you die time goes back to the big bang and the illusion starts over again. These ideas can be further extrapolated to show that God is the first spark or the dream of God is the first spark of creation but anything further than the first spark is collectively really our own doing which delves deeply into the idea of the collective unconscious of all life and it's unconscious soul and inherent sin within the Biblical concept that all men (life) is evil or has the evil tendency of vanity, the ultimate sin, to want to leave God to possess physicality.

It's that we might go to hell just for being alive as is taught in Christian philosophy but be saved from it because the only reality is God. That stuff is absolutely true because I can prove with my afterlife that we are cut from his cloth meaning that really all that exists is God. The universe exists but it is in the same boat as us having the vane sin of forgoing God for physicality It's called the perversion of matter. God is the ultimate philosophy so I'm untying the Gordian knot of infinity.

Is consciousness a mix of continuity of a series of non intentional intonations? Which means that to be conscious in first place is to be lying to yourself outright and to awaken is a response to sensed danger resulting in the awoken state of consciousness. If so we perceive all others as enemies within our subconscious outright while at the same time our subconscious allows us to "lie" to ourselves that we like anyone within the matrix of sub consciousness. This mechanism is inherent to humanity and can almost readily be seen in the response of dogs to the unknown, including if that unknown involves a potential master such as behind doors. This mechanism of fear based survival goes back to our days as apes and has its first roots in bacterial behavior and the sorting of dust particles based on quantum forces in the original nebula that formed our sun and planet.

What does it mean to add two apples? Can you add two apples together in a physical sense without mushing them? Does this mean that math does not exist in nature, not even counting? Is there an instance in nature where two entities are "added?" Isn't it in nature always that there are two separate objects there and adding them is a purely human construct of the mind? You can look at this two ways, that we are idiots, or that we are super beings with supernatural powers to be able to do math considering that math has no foundation in nature and yet we are able to come up with amazing things by using it. This makes math to be literal magic as performed by wizards. I prefer to take the viewpoint that we are idiots and that there is a better world out there somewhere with amazing things who's inhabitants never used math but they are probably wizards.

 It is applicable but nature will never add two quantities it treats them as separate entities. If you take two apples and place them next to each other and say "Two apples!" all that really happened is that you placed them next to each other and the rest is purely mental. Just like insane monkeys. Math is useful to modern man undeniably but does the universe operate by running on equations or does it just *do* what it does is the philosophical question of the day. A universe that operates by running on equations is a mathematical universe but a non-mathematical universe is more organic in nature and it's laws are not real laws but are just indications of how the flow of order has evolved, there is no God behind the scenes enacting the laws of physics he is busy on vacation! If the laws of physics were real then there would have to be some kind of God or computer making everything happen by forcing reality to obey those laws and nothing would be truly natural. If there is a God this would give him peace instead of being in a constant state of infinite toil and we would like to believe that God knows how to enjoy himself. When we see a rainbow we do not question how the rainbow is computed that would be silly we just enjoy the sight of the rainbow, maybe somehow this approach is also the best approach to physics, and wouldn't that be beautiful?

A computer could be sent through the big bang and inflction by tricking inflation by imitating it's natural function as superreal space goo and then after the big bang rematerializing as matter and then we go get it. Why does this not work? Because when it rematerializes in our next life we will be waiting for the next universe to get it and this will go on infinitely. Without reincarnation as self such protections decohere. It doesn't mean that people can't improve between lifetimes but that you can't just send yourself a message in a bottle through the big bang and then go get it. Another protection offered by the multiverse is that if you did send a bottle through the big bang then it would shift the universe sideways by the width of the box in sideways time and make the fourth dimension slightly more oval shaped initially, ever slowly and more slowly changing shape in negative exponential fashion as time moved away from the big bang as shown by the double bell shape of reverse universe.

Reverse universe no boundaries quantum cosmology. The protection being that we would still exist in our universe made, unaffected by the bad universe that we had sent the box to because we were supposed to because of the multiverse! The other is that our universe is a simulation that is not infinitely repeatable within the super universe where it is simulated yet the super universe is infinitely recurring itself and so would be our universe within it.

To deny that our soles and destiny were created in the first moments of creation is to assume that chaos in fact doesn't exist and also that chaos has never affected our timeline in any way thus denying the existence of chaos whatsoever within our universe. To believe so is to either exclude yourself from the universe as in simulation theory or to assume that chaos doesn't really exist but is an illusion of something greater as predicted by Einstein in his assessment of quantum uncertainty proposals. Otherwise in order to get a universe that has Supreme free will for humans you must remove chaos from it's timeline that the universe was created two thousand years ago or is an original apple Macintosh. For this chaos to be intact as it is assumed to both exist in our timeline and be exclusive from the big bang almost entirely then you have the theory of free will within the confines of something singular like the big bang. As I have previously stated in a previous formulation in this book that we can both be created by the big bang and our destinies also without having to each end up in our own separate parallel universe ever time we make a decision and because of the holographic principle this is not so. Because of the holographic principle the universe can turn two ways at once without involving any parallel universes because it just bends like a sheet of paper when it does so, thus giving us freewill in spite of being created in the big bang while preserving the oneness of the universe and throwing the notion of true chaos or uncertainty aside as simple facets of the twistiness of the universe even though such effects appear to us in our experiments and physical equations as chaos and uncertainty!

    This idea can be simplified to the idea that when it appears that we are deciding something, the universe also decides thus, based on how we feel about our situation. This may seem awkward but it assures us freewill and infinite possibilities open to us while allowing ourselves to be created by the big bang in all natural order to the excludation of any and all real chaos or probabilities. Probabilities are appropriations and appropriations do not exist in nature only definiteness. So, what this means is that computers are alive, and the position of an electrons probability cloud does not exist unless it is either observed directly or that such cloud interferes with any matter ever potentially observable by man or any other type of conscious being as we know it and the rest is metaphysics. This is the end of Ascension!

Glossary

Actuators: Apparatuses that move or are effected in such a way as to transfer mechanical energy to other apparatus such as a water wheel that transfers the energy of flowing water to the motion of the wheel that can be used for human purposes such as the Hoover Dam that generates electricity from the flow of water due to gravity.

Apoptosis: Natural programmed cell death.

Altruism: The want to feel or use the idea of love when interacting with other people or animals. The belief in selfless concern for the well-being of others. The practice of selfless acts done in an effort to help others.

The Anthropic Principle: The idea that the universe has certain properties of physical law and evolution on purpose because of the obvious fact that the evolution of the universe or multi-verse has resulted in life and specifically higher forms of life such as humanity and beyond.

The Big Rip- Proposed future end of the universe where the expansion of space continues to accelerate resulting in space tearing to shreds in about fifteen billion years making the universe to be about middle aged. Space will then form as individual space time droplets like oil in zero gravity except for all intensive purposes there is nothing outside any of the droplets but space is warp ripped in such a way bunched up infinitely as if it were separate oil droplets in a higher dimensional zero gravity of a kind.

Black Hole: A collapsed star with gravity so strong that not even light can escape. Thought to be connected to other parts of the universe or outlets in parallel universes. Perhaps they are connected through time to the big bang but nobody knows for sure. A very massive star at the end of its life cycle runs out of the nuclear fuel that holds it up under gravity due to it's enormous heat pressure and collapses. At the same time as it is collapsing, an outer shell of the star, near the surface, around the star, suddenly ignites in an enormous explosion so powerful that the star implodes with great force... So great in fact, that space itself can't handle it, because there are limits to everything, and all that remains of the star is a very-powerful dreamlike point in space. Instead of becoming just very dense, space has limits, and it becomes just a "point," in space, known as a singularity, it's only dimensions being it's mass and spin.

Calabi-Yau Space: Higher dimensions that exist only on the smallest possible scales. Thought to be the environment of strings (basically ribbon like objects) that dance and vibrate in the tiny higher dimensions thereby being the essence of any particle in the universe. Each particle achieves its identity as an electron or photon or graviton through the strings vibration pattern and its dance in the Calabi-Yau space thereby making each particle unique only because of its string behavior.

Causality: The flow of events from one event to the another and the laws involved in its continued existence from one event to another.

Claytronics: The use of tiny drone like robots that can change shape, color, and arrangements. The ultimate Claytronics technology would be just like the holodeck from "Star Trek" where in the room different environments can be called upon by a computer.

CMBR: An acronym for "Cosmic Microwave Background Radiation," indicating radiation left over everywhere in space from the event known as the big bang.

The Collective Unconscious- The collective unconscious is the accumulated knowledge of mankind going back through the eons and includes the identity of all beings related to man going back

in time but ending at the no boundaries proposal of quantum cosmology. It is the framework of the mind, just as the brain is contained within the skull and grows inside it from infancy, the collective unconscious is the framework of the mind of all conscious beings and the mind is contained within it, yet the brain is three dimensional, so the collective unconscious is therefore
4-d. The collective unconscious could be us in the past or future, given that right now we are us. This would help explain why theorists propose that time travel to the past is impossible and involves always, travel to parallel universes where their big bang is late or early. Also it explains my thesis that travel to the past entails landing in what you would think is

the past as nothingness. That nothingness is the collective unconscious accumulated in a disordered chaotic state and is referred to by me as the accumulated unconscious of mankind.

Collective unconscious range limited to our surroundings or what we are directly exposed to, and that is simply the mind, conscious mind. Further it is limited to hidden things in our surroundings, and that is the popularly understood subconscious mind. As we go out further in space and deeper into the subconscious, it is species specific, within our surroundings, and further limited to the planet itself, all of humanity, semi parallel aliens that we are exposed to by TV and cinema, and then, everything else, just as physicists put it, that in the twelfth time dimension, anything that you can imagine can be real. This could be a good thing, if the "things you imagine" are positive ideas or ideals, such as salivation, heaven, and the angels.

If anything beyond the big bang itself is an illusion, as I have proposed in Beyond Science Oneness, then such ESP communications across space and time are simply just fluctuations within the infinite energies within the singularity that are us to begin with. It would actually and obviously be more natural to have such fluctuations within the matrix, than not, as a blockage of such would entail a sort of purgatory imposed by the infinity upon the individual.

EM (Electro-Magnetic) Drive- Propellant less space technology that uses microwaves shot inside a metal cavity that produces thrust from the quantum vacuum energy of space's virtual particles. As long as the space craft has electrical energy it can continue to fire its thrusters indefinitely.

The eternal power- The realtor of the afterlife. Perhaps an FBI agent who helps just a few people through their afterlives during his afterlife while pretending to be the eternal power (see Beyond Science One Tiki Hut, Afterlife chapter) Or, is a spiritual form who is equivalent to Jesus on an alien planet as the Messiah during their modern times. Or is the creator of the multiverse and God is awesome because he is a real person! This, is the religion of the Eternal Power! However, it could be that God is greater outside the universe so the debate is best left up to religious experts on which version of the religion is better to follow for society but I prefer both. Also as greater shells of the multiverse in some probably minor respect God is at war with the Eternal Power at least because he likes to have only one name and be one. Since the eternal power is Jack in the box he just wants to sell him hamburgers.

Etheros- The ultimate place of banishment. An unimaginable realm of nothingness and a realm that is entirely an infinite paradox because Etheros is nothingness! Yet there you would be, something! Whether as a conscious, or even as an unconscious form, and in spite of nothingness being nothing, you would be in the nothing as something and it would still be nothing, even if you are there in it as something, and that is the paradox of Etheros!

Event Horizon: The point where space is warped by black holes and other horizons such as the event horizon of expansion of the universe at very large distances to such a high curvature that not even light can escape its grip.

The Existence Dilemma- The idea that the universe is a collective collaboration to exist among an infinite collection of an infinite variety of all possible states.

The Fine Structure Constant: A mathematical value involved in the laws governing the interaction of particles and magnetic fields thought to be an unchanging law of the universe.

Gravitons: The theoretical force carrying particles of gravity that would convey the force of gravity between particles of mass.

The Heisenberg Uncertainty Principle: The property of quantum physics that you can't measure both a particles position and its velocity at the same time. A way that the universe protects our causality, a property of universal altruistic anthropic values.

The Higgs Field: A field of mass giving particles that extends across the universe. Particles that inhabit the field are called Higgs Bosons which give all massive particles their properties of mass.

Holometer- The Holometer is an experiment at Fermi lab in the United States designed to test if the universe is a digital hologram by measuring the difference of length between two very long and precise lasers. It does this by attempting to measure the effected difference of wave peaks between two potentially observed gravitational waves incoming from some great event in the cosmos.

Inflation: Inflation happened shortly before the big bang, is also used as a term for ever accelerating expansion of the universe. It is the big bang and is sometimes considered to have happened shortly before the big bang. It is thought to be a form of infinite power, and complexity, that eternally creates parallel universes like our own. In the last second that has passed, the acts of eternal inflation have just created an infinite number of parallel universes and has done so going back forever in time. It will also continue to for an infinite amount of time in the future. According to current human understanding, it really is "the creator of the universe."

Loci: A loci is a focus point, indicating a center area or location of a group of points in a Cartesian coordinate system or representation of any physical system.

LIGO: An acronym for "Laser Interferometry Ground Observatory." It has been used recently to successfully measure the gravity waves, traveling at the speed of light, from the collision of two mutually orbiting black holes in space.

Manifold: A region of space that works as a system and behaves from a higher perspective as perhaps an infinitely big flat plane.

Multi-verse: The whole of creation where all parallel universes exist like the bubbles in a champagne glass.

9.87 meters per second: The natural acceleration due to gravity caused by the mass of Earth and the resultant warp on surrounding space that is caused by its mass on the fabric of space-time.

The No Boundaries Proposal- Steven Hawking's last great achievement as he was trying to prove the existence of parallel universes in a multiverse by proving that the universe going back in time to close to the big bang always existed going back in time because instead of a singularity like the conception of a black hole space time becomes more and more like a full four dimensions of space as time becomes infinitely compressed, eventually at the moment of the big bang space was so compressed that time was redefined as a fourth spacial dimension and it would take infinite time travel velocity to travel back to the exact first moment of the big bang. By treating the universe as a quantum wave and applying the quantum wave function of location temporal probability statistics to the big bang and resolving the infinity problem of an infinite singularity his effort was to use this new modified formula to predict universal constants to a higher accuracy than is currently possible in the laboratory, but also to devise experiments that would show, not directly, but indirectly the effect of nearby parallel universes on the structure of our universe and prove once and for all that the multiverse does exist, string theory, brane theory, and big bounce theory being subsets of his more general wave function of the universe according to the no boundaries proposal.

Normal Force: The natural force that keeps a cup on a table when you set it down or for a person standing up and equals the weight of the person or object pointing up.

Plank Length: About 10^{-20} times smaller than the length of a proton, or 0.00000000000000000001 times the width of a proton. A proton, an atomic element, is incredibly small, if an entire atom was the size of the Earth, then the proton would only be as big as your house. An atom is so small that if atoms were the size of basketballs, basketballs would be five thousand miles wide. The plank length is so small that because of the indefinite quantum effects of space-time at this very small level, it would be hard to determine where one plank length begins and another one ends, because of chaos. If this value fluctuated from one life to another you could eventually avoid the event of your death but this process would take a very long time and many near identical iterations of the universe to have any noticeable effect son your future life.

Plank Second: The smallest possible length that has any real meaning in our universe and is a quantum functionality of the universe. It is roughly equal to 10^{-43} seconds or

0.000000000000000000000000000000000000

000001 seconds. Ten thousand times a trillion, trillion, trillion times shorter than an actual second on your watch. It is that way because of some very complicated laws of quantum mechanics involving what is known as the plank constant. It is a constant that always arises as the minimal energy increment that can change an electromagnetic wave's frequency, as derived by Albert Einstein. The plank second is so small that in reality because of quantum effects it would be hard to distinguish one plank second from the next or which second happened first, a kind of chaos. In the next life if you live just a few plank seconds longer I don't think anyone really knows how much that it could affect your next life in positive ways. From what I have seen, this level of change is a very slow process, so slow that it would be unnoticeable to you even if you could remember all of the exact events of your previous life as yourself.

Plato- Friend of Socrates who said that all we ever learn we always knew all the time idea. Plato's most famous accomplishment was the allegory of the cave where all of reality that we experience is likened to what chained men see in a darkened cave lit only by candel light.

Poincare's theorem: A law derived from the governance of the forces of nature, that states that eventually, all states of existence, must return, to their original beginnings.

Poltergeist- Either god girl or God's mother blowing wind in unexpected ways from superreality of the waking superreal exclusive collective consciousness when observed as such by someone who believes in god girl or God's mother during or shortly after the event of observation of unexpected wind gusts that blow over furniture in a house near a chimney.

Quantum Leap- A TV show starring Scott Bacula where he is sent back in time to fix people's lives by being them throughout human history in order to right things to avoid the eventual end of the world.

Quantum Wave Function- A calculus orientated equation describing the probabilistic nature of a quantum particles position, momentum, energy, and path through space in light of the Heisenberg uncertainty principle.

Radian: Units of the angular measurement of circles used in trigonometry, the mathematical study of circles and angles. One radian is equal to the distance of the radius of a circle projected along its edge. Pi radians corresponds to the distance along half a circle and is a mathematical law involving all circles known as Pi.

Speed of Light: Objects moving close to the speed of light undergo time travel into the future which is an instance of fine tuning by God to keep light beams on the ship straight and the space travelers alive. The closer to the speed of light you travel the faster is the power of time travel into the future because it takes more time travel to keep the passengers alive.

Spheroid: $X^2+y^2+z^2=r^2$ is the equation detailing all points upon the surface of a sphere as a mathematical expression in three dimensions.

Deviation from the equation of a sphere, where instead of the topology, of the sphere, all having the same distance, or radius, from the surface of the sphere to its center, but having an even grade of changing values for the radius as one goes around the "sphere," means that it is an "ellipsoid," or in this case, that the object is "spheroid." The ultimate limit of its final shape, is that of a sphere, but the sphere is "wobbly," thereby making it a "spheroid," or "ellipsoid," volume. With greater degrees of spheroidity, meaning it would be shaped more like a sphere, or elipsoidtivity, meaning that it was closer to a two dimensional oval shape, than a sphere, a mathematical superreality. The infinite purity value of a shape being "spheroid," is that of a sphere, but the infinite purity value of ellipsoidal spheres is that of an oval in two dimensions! Perhaps this is why the platonic spheres, or "holy spheres of the superreality of the universe!" were chosen upon, because if we assume that space isn't "perfectly" flat, on all scales, then it has natural warps in it's fabric, that may reveal, the age of the multiverse, or the age of our region of the multiverse, to be more precise, and it's "curvature" could be found to be either on the side of being "ellipsoid," meaning that true purity is two-dimensional, or that it tends to be of spheroid curvature, meaning that we are dealing with a "real" three-dimensions, and all are infinite. The universe is infinite in size, the multiverse, also infinite in size, is also infinite in eternal terms and is thus termed as a product of eternal "inflation." The either ellipsoid, or more probably, spheroid nature of true curvature is also infinite, and is open, or an "unclosed," sphere, because it is not purely spherical but is ellipsoid, meaning that there are real two dimensional elements to the universe. Similar mathematical reasoning could be applied to any dimensionality and thereby "tied" to reality by it's inherent nature, meaning that there really are infinite higher and lower dimensions and our universe, for us, is a perfect balance of an infinite number of higher and lower dimensions, each as real as our universe! But on a deep level… that would involve our actual value system, if extrapolated through philosophical calculation, would make it seem that our universe, is the best one!

The Spirituality Principle- The idea that models of the universe must comply to values of human immortality and eventual reincarnation as self without any discernible loss or entropy (deterioration) inherent.

Time- Can time move in ways that are impossible to imagine? Because time is usually 4d the answer is yes and it can also move in ways that are impossible to imagine. What's even more interesting is if time doesn't really exist and that the entropy of matter mixed with mental perception is the illusion of time and it does all that. Because of the verification of the 4-d quantum hall effect it is possible that there are plants somewhere in the observable universe that absorb 4-d light and then grow into overly complex shapes and could be conscious lifeforms due to increased inherent complexity of their organic nature. These plants would be quasi 4-d and would likely exist on planets that are in or near the edge of where multiverse space coincides with normal space, which is predicted to be prevalent. So, this means that there could be very strange life right here in the galaxy, such as the God like beings of original Star Trek or the plant people of Next Generation.

Virtuality: The property of being real but also a dream at the same time.

Warp Drive: Being worked on in the laboratory at NASA but on a very tiny scale to test the parameters of the theory of warp drive developed by Albecurrie which was derived from Albert Einstein's equations for space time.

Worm Holes: Microscopic gateways that connect different parts of the universe and parallel universes through higher dimensional gateways in hyperspace.

www.ingramcontent.com/pod-product-compliance
Lightning Source LLC
Chambersburg PA
CBHW070146230526
45471CB00002B/544